AUVSI XPONENTIAL 2021

AA006245

Online
4 – 6 May 2021

Atlanta, Georgia, USA
16 – 19 August 2021

ISBN: 978-1-7138-5222-3

Printed from e-media with permission by:

Curran Associates, Inc.
57 Morehouse Lane
Red Hook, NY 12571

Some format issues inherent in the e-media version may also appear in this print version.

Association for Unmanned Vehicle Systems International
2700 South Quincy Street
Suite 400
Arlington, VA 22206
USA

Phone: (703) 845-9671
Fax: (703) 845-9679

www.auvsi.org

Additional copies of this publication are available from:

Curran Associates, Inc.
57 Morehouse Lane
Red Hook, NY 12571 USA
Phone: 845-758-0400
Fax: 845-758-2633
Email: curran@proceedings.com
Web: www.proceedings.com

TABLE OF CONTENTS

INTELLIGENT REAL-TIME SCHEDULING, DISPATCHING AND MONITORING SYSTEM FOR UNMANNED VEHICLES

George Rzevski

A considerable amount of research effort is focused on unmanned vehicle on-board intelligent technology and not enough is done to ensure that unmanned aircraft and road vehicles are intelligently scheduled and dispatched and then monitored during operation. This paper reports on the author's research into requirements for scheduling, dispatching and monitoring unmanned aircraft and road vehicles and his novel architecture for a system which meets these requirements. The proposed system is based on complex adaptive software technology and features emergent digital intelligence.

INTRODUCTION

At present, vehicles (including land vehicles and aircraft of all types) are scheduled, dispatched and monitored by human dispatchers aided by a variety of packages. Our research identified considerable inefficiencies and waste of resources associated with current transportation management. In author's view, it is essential to develop unmanned Intelligent Adaptive Systems for scheduling, dispatching and monitoring of transport and, in particular, for unmanned transport.

STRATEGIC IMPORTANCE

Current focus on investing into *production* of spacecraft, electric aircraft, electric cars, electric trucks, electric drones and flying cars, is not matched by the investment into systems which will be required to *manage* all these sophisticated flying and land travelling vehicles.

There is an urgent need to remedy this situation and channel funds into the development of autonomous intelligent fleet management systems. Without advanced fleet management systems new aircraft and vehicles will not be optimally organized, scheduled and utilized.

REQUIREMENTS

To avoid generality, let's describe requirements for a representative example - air taxi operation. Requirements for other transportation modes will be similar and yet different in details or scale.

To manage air taxi operation, manned or unmanned, it is necessary to perform the following activities:

- Receiving requests for seats
- Accessing all relevant data (databases, data streams)
- Creating flights
- Allocating appropriate physical, human, financial and knowledge resources to newly created flight

----

Professor, owner of Rzevski Research Ltd, 3 Ashbourne Close, London W5 3EF

- Calculating costs of the new flight
- Biding for seat requests
- Dispatching and monitoring flights
- Rapidly rescheduling affected resources if any unpredictable disruptive event occurs

Unpredictable disruptive events may be desirable or undesirable and include unexpected seat requests, cancellation or modification of seat requests, aircraft failures, nonavailability of crew members, change of weather conditions, etc.

A SOLUTION

To meet requirements and avoid inefficiencies and waste evidenced in current practices, the author suggests that we need *Unmanned Intelligent Adaptive Scheduling, Dispatching and Monitoring* of flights and other transportation missions.

This is particularly important in cases of managing unmanned vehicles, where the margins of acceptable management errors are very small.

WHY UNMANNED?

- There are not many skilled dispatchers around to meet a seriously growing demand, and they are expensive
- Human operators make mistakes, especially under pressure
- Data required for decision making is dispersed and not easily accessible
- The speed required for making a dispatching decision in cases of disruptions or emergencies is often beyond human abilities

WHY INTELLIGENT?

Scheduling and dispatching decisions are rarely straightforward – there is often an uncertainty which option is the best.

Intelligence eliminates, or at least reduces, uncertainty of outcome by applying domain knowledge.

For scheduling, dispatching and monitoring the following knowledge is required:

- Common-sense knowledge (e.g., aircraft fly above the ground)
- Theoretical knowledge (e.g., time to reach a destination = distance x speed)
- Experience knowledge (e.g., check pilot's license validity before scheduling)

Human intelligence (HI) is excellent in applying common-sense knowledge and experience-based knowledge gained by working for particular transport operator. Also, human intelligence is superior in strategic decision making where excellent understanding of the whole context of business, economics and politics is required. Applying theoretical knowledge is usually delegated to packages.

Artificial intelligence (AI) is excellent in extracting knowledge from data, knowledge that is equivalent to experience-based knowledge because data captures past performance. Data also captures common-sense knowledge, albeit only that part, which was practiced by the operator (e.g., data collected by dispatching aircraft for airlines cannot be used for dispatching of air taxis or flying cars).

Traditional AI, such as artificial neural networks trained on data, lacks theoretical knowledge and is therefore not able to help with dealing with disruptive events that were not captured by data used in training. As a consequence, traditional AI is *not adaptable*.

In searching for adaptable AI, the author has developed *Complex Adaptive Software Technology* (CAST) which uses all three variants of knowledge stored in a *knowledgebase* and is superior in operational decision making – it is precise, reliable and rapid.

WHY ADAPTIVE?

Under complex operating conditions there are frequent unpredictable disruptive events as described above.

Adaptive systems are capable of eliminating, or at least reducing undesirable consequences of disruptions by rapidly rescheduling affected resources.

Here is how adaptability works:

- All relevant data sources are continuously monitored
- If an unexpected event occurs, consequences are rapidly assessed and affected resources are rescheduled to eliminate, or at least reduce undesirable concerns

Since unpredictable disruptive events tend to occur quite frequently, the rescheduling of affected resources has to be done in *real time* – before the next disruption.

FUNCTIONS

Let's again avoid generalities and focus on air taxi operation.

An Unmanned Intelligent Adaptable Scheduling, Dispatching and Monitoring System for air taxi operation must be able to perform the following functions:

- Receiving request for seats via user-friendly booking system capable of autonomously discussing with the customer flight options and negotiating seat price
- Accessing all relevant databases (aircraft readiness, pilot experience & training/licensing records, airport suitability, weather go/no go)
- Creating a new flight
- Selecting aircraft for the created flight considering flight-readiness, cost of repositioning and any other criteria specified by the client
- Selecting crew for the created flight using criteria such as license validity, rest time, cost of travel to aircraft, and any other criteria specified by the client
- Calculating cost of the created flight, including repositioning of the selected aircraft, cost of crew travel, cost of using airports, cost of fuel and overheads
- Sending calculated seat prices to the booking system within few minutes of a seat request
- Scheduling aircraft maintenance
- Searching for the most appropriate nearest repair station and booking repairs, if aircraft fails
- Continuously monitoring critical data sources and instantly detecting any disruptive event
- Within seconds, identifying which part of the operation will be affected
- Rapidly rescheduling affected parts of the operation to eliminate the consequences of disruption, always maximizing the enterprise value
- Rapidly calculating costs of disruption and accordingly adjusting transportation costs
- During intervals between disruptions, analyzing previously agreed schedules and costs, and, if necessary, making corrections or improvements

ACCOUNTABILITY

The unmanned dispatcher must have a dashboard enabling human dispatchers to monitor the system decisions and to change them, if necessary. The system must also provide feedback to human dispatchers on the effectiveness of their interventions.

BENEFITS

Based on 20 years of experience in researching and developing Complex Adaptive Software Technology, the author arrived at conclusion that an Unmanned Scheduling, Dispatching and Monitoring System based on this technology would offer the following benefits.

Uniqueness

To the best of author's knowledge, there is no system or package on the market that can compete in terms of coverage, intelligence, adaptability and price.

Improved Quality of Service

Unmanned dispatchers would reduce delays and cancellations by rapidly rescheduling affected resources whenever a disruptive event occurs and by thoroughly checking agreed schedules and costs of transportation in intervals between disruptive events.

Cost

CAS technology is considerably less costly than traditional AI. Dispatching systems based on this technology would cost less than current software packages for dispatching and could be available as SaaS.

Profitability

By improving utilization of resources and removing undesirable effects of disruptive events, operational costs would be reduced by at least 20%.

Return on Investment

Similar systems applied to different applications repaid for themselves in 6 months, when purchased on license.

Transparency

Costing transparency would be achieved by calculating costs of every individual resource deployment, down to individual transactions.

Productivity

An estimated 50% of jobs in transport operations could be replaced by complex adaptive software, substantially increasing productivity. Complex adaptive software operates 24 hours a day, 7 days a week, continuously updating schedules in reaction to disruptive events or on improving performance.

Future Proof

Complex adaptive software can be easily updated or upscaled and its widespread application would create new jobs related to design, coding, commissioning, maintenance and repair.

Improved Management

Unmanned dispatchers take over a very large management load by autonomously making all routine scheduling decisions and thus enabling managers to focus on strategic and tactical issues.

UNMANNED DISPATCHER ARCHITECTURE

The key elements are

1. Knowledgebase, where domain knowledge is stored
2. Digital World in which Digital Agents exchange messages and negotiate solutions
3. Interfaces between Digital World of the dispatcher and Real World in which business operate

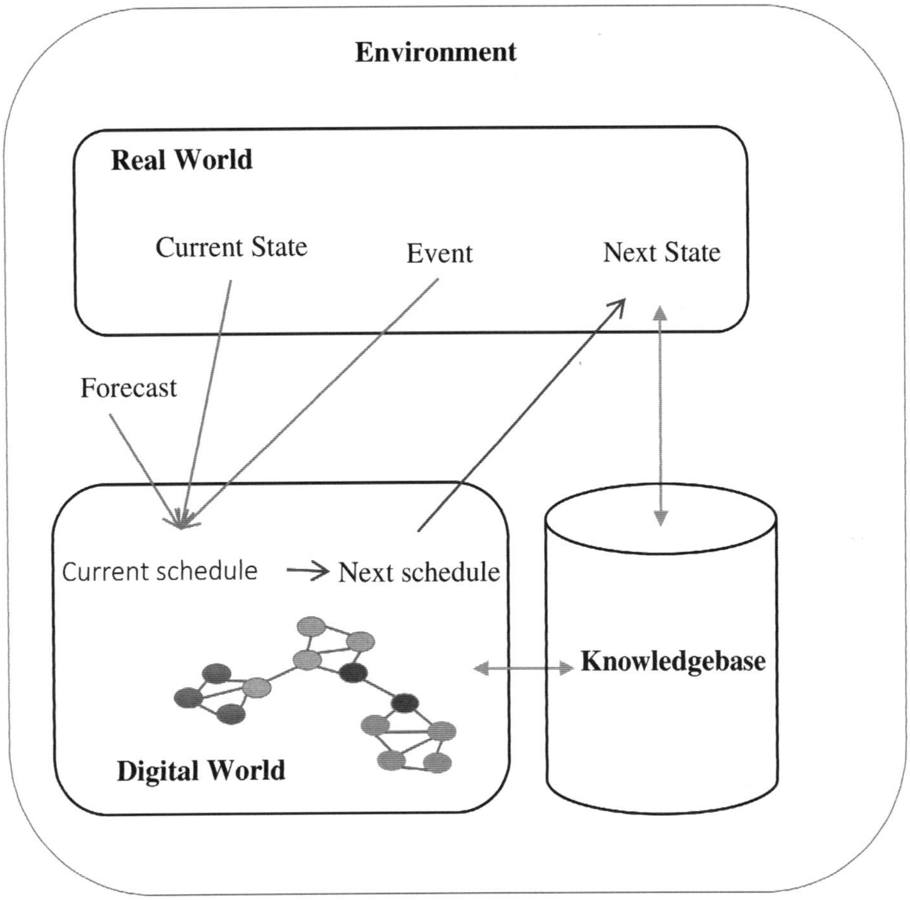

Fig. 1. Architecture

Knowledgebase

Consists of *ontology* and *data*. Ontology includes *object classes*, *object classes relations*, *object classes properties* and *agent scripts. Object classes* include: Customer, Seat, Flight, Airport, Runway, Airplane, Pilot.

Digital World

Digital world is the world of digital agents. Agents create schedules, dispatch fleets and calculate costs of flights and price of seats in the Digital World. An example how digital agents work is given below.

- A seat is requested on a flight between Airport 1 and Airport 2 for a particular date and time
- Client12Agent is assigned to the client who requested a seat
- Flight12 is created
- Flight12Agent is assigned to Flight12
- Flight12Agent sends messages to Aircraft Agents and PilotAgents asking which aircraft and which pilots are available for Flight12
- Agents of available aircraft and pilots send their bids to Flight12Agent including cost of repositioning and travel in time for the Flight12
- Flight12Agent selects aircraft and a pilot for the Flight12
- Flight12Agent asks CostService to calculate Flight12 costs and sends projected seat price to Client12Agent
- Client12Agent negotiates seat price with the prospective client; the agent is allowed to offer certain discount, if necessary

Interfaces

Digital World monitors data generated by the Real World - demand forecasts, current state of the Real World, and disruptive events.

Real World receives schedules and seat prices from the Digital World.

COMPLEXITY SCIENCE

The subject of complexity science is the behavior of complex systems, that is, systems which consist of a large number of autonomous components, called agents, engaged in intense interactions. Overall behavior of complex systems emerges from the interaction of agents and is therefore unpredictable although not random – it follows discernable patterns.

Complex Systems are usually called *Complex Adaptive Systems* to emphasize their key property – the ability to selforganize and adapt to changes in their environment.

Complexity Science is a new science, developed in the 1990s primarily at the Free University of Brussels by Prigogine[1, 2] and at the Santa Fe Institute by Kaufman[3] and Holland[4].

The author's contribution is experimental – his team builds large-scale complex adaptive systems for commercial clients and deduces scientific principles of complexity from their behavior[5]. The building blocks for Author's systems are agents – short algorithms that exchange messages with each other. Problems are solved and conflicts resolved by agent negotiations. Before acting, agents always consult knowledgebase where domain knowledge is stored in computer readable format.

REFERENCES

[1] Prigogine, Ilya, "The End of Certainty: Time, Chaos and the new Laws of Nature". Free Press, 1997.

[2] Prigogine, Ilya, "Is Future Given?" World Scientific Publishing Co., 2003.

[3] Kaufman, S., "At Home In the Universe: The Search for the Laws of Self-Organization and Complexity". Oxford Press. 1995.

[4] Holland, J. H., "Hidden Order: How Adaptation Builds Complexity". Addison Wesley. 1995.

[5] Rzevski, G., P. Skobelev, "Managing Complexity". WIT Press, Southampton, Boston, 2014. ISBN 978-1-84564-936-4.

NOTES

NOTES

NOTES

NOTES

NOTES

NOTES

NOTES

NOTES

NOTES

NOTES